Bizarre Beast Battles

Scorpion vs. Black Widow

By Caitie McAneney

Please visit our website, www.garethstevens.com. For a free color catalog of all our high-quality books, call toll free 1-800-542-2595 or fax 1-877-542-2596.

Library of Congress Cataloging-in-Publication Data

McAneney, Caitie.
Scorpion vs. black widow / by Caitie McAneney.
p. cm. — (Bizarre beast battles)
Includes index.
ISBN 978-1-4824-2792-9 (pbk.)
ISBN 978-1-4824-2793-6 (6 pack)
ISBN 978-1-4824-2794-3 (library binding)
1. Scorpions — Juvenile literature. 2. Black widow spider — Juvenile literature. I. McAneney, Caitie. II. Title.
QL458.7 M384 2016
595.4'6—d23

First Edition

Published in 2016 by
Gareth Stevens Publishing
111 East 14th Street, Suite 349
New York, NY 10003

Designer: Katelyn E. Reynolds
Editor: Therese Shea

Photo credits: Cover, p. 1 (scorpion) efendy/Shutterstock.com; cover, p. 1 (black widow) Matteo photos/Shutterstock.com; cover, pp. 1–24 (background texture) Apostrophe/Shutterstock.com; pp. 4–21 (scorpion icon) mr.Timmi/Shutterstock.com; pp. 4–21 (black widow icon) SFerdon/Shutterstock.com; pp. 4, 14 2happy/Shutterstock.com; p. 5 Sukpaiboonwat/Shutterstock.com; pp. 6, 9 (black widow), 13, 17 Peter Waters/Shutterstock.com; p. 7 James H Robinson/Science Source/Getty Images; pp. 8, 9 (map) ekler/Shutterstock.com; p. 8 (scorpion) wacpan/Shutterstock.com; p. 10 DON EMMERT/AFP/Getty Images; p. 11 Glenn Asakawa/The Denver Post/Getty Images; p. 12 Amound Quanjer/Shutterstock.com; p. 15 DSTU/Shutterstock.com; p. 16 Audrey Snider-Bell/Shutterstock.com; pp. 18, 19 James C. Bartholomew/Shutterstock.com; p. 21 (scorpion) Mauro Rodrigues/Shutterstock.com; p. 21 (black widow) Nate Allred/Shutterstock.com.

Printed in the United States of America

CPSIA compliance information: Batch #CS15GS: For further information contact Gareth Stevens, New York, New York at 1-800-542-2595.

CONTENTS

Words in the glossary appear in **bold** type the first time they are used in the text.

SCARY SCORPIONS

Scorpions and spiders might not look related, but they both belong to the arachnid family. Arachnids have bodies with two main segments, or parts, and four pairs of legs.

There are around 2,000 species, or kinds, of scorpions! Scorpions are strong and tough. They can survive harsh weather and attacks from other creatures. They're **carnivores**, but can live on just one insect per year when food is hard to find!

These creepy critters have a stinger at the end of their tail that holds strong **venom**.

Scorpions can survive extreme heat and cold. Scientists have frozen scorpions overnight. Then, they unfroze the scorpions, and the scorpions started moving again!

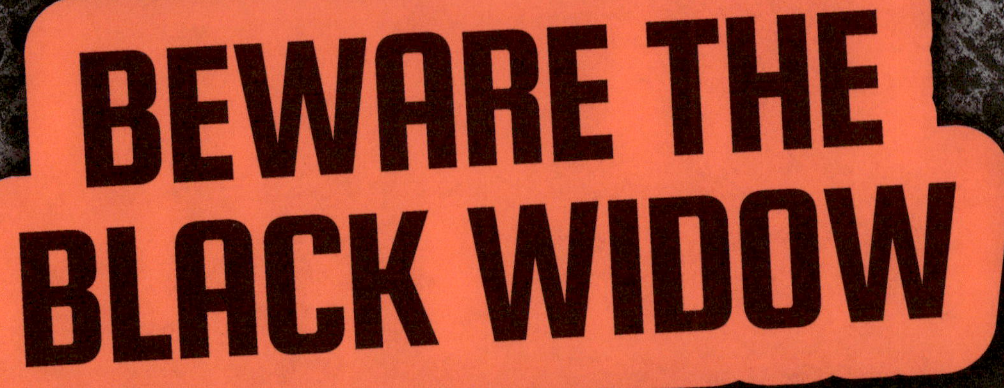

BEWARE THE BLACK WIDOW

Black widow spiders are some of the most feared arachnids in the world. Their thick, black body and red markings make them easy to spot. Black widows are part of the *Latrodectus* group, which includes more than 30 species.

Latrodectus is Greek for "biting in secret." These spiders don't think twice about taking a bite if they're hungry, trapped, or in a fight. They'll even take a bite out of their own kind. Luckily, these web-dwelling carnivores are mostly interested in eating insects.

Female black widow spiders (bottom) are much larger than males (top), and their venom is much stronger.

BATTLE OF THE ARACHNIDS

Could these two arachnids ever run into each other? Let's look at their **habitats** to find out! Scorpions are known for living in deserts. However, they can also live in rainforests and mountain regions.

WARM AND DRY REGIONS

scorpion habitat

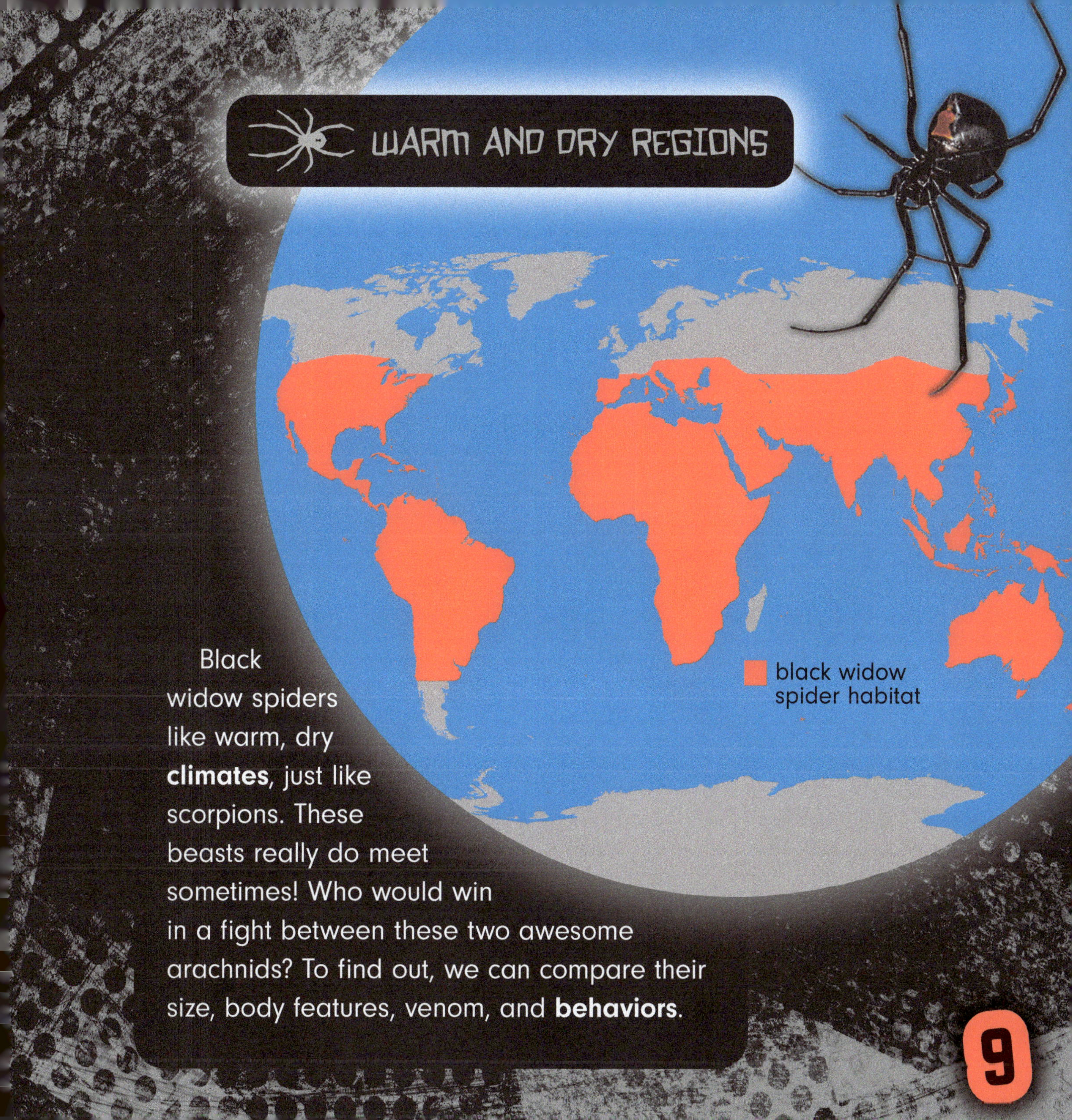

WARM AND DRY REGIONS

Black widow spiders like warm, dry **climates**, just like scorpions. These beasts really do meet sometimes! Who would win in a fight between these two awesome arachnids? To find out, we can compare their size, body features, venom, and **behaviors**.

In both scorpion and black widow spider species, females are bigger than males. Scorpion size mostly depends on species. The Middle Eastern scorpion is only 0.25 inch (6 mm) long, while the flat-rock scorpion grows to be more than 8 inches (20 cm)!

WEIGHT:
UP TO 2 OUNCES (57 G)

LENGTH:
UP TO 8.3 INCHES (21.1 CM)

Female black widow spiders are double the size of males! However, all black widows are small and light. So, scorpions are usually larger than black widows. However, size might not be that important in this fight. Both scorpions and black widows have some mighty **defenses** as well as fearsome **weapons**.

AWESOME DEFENSES

What are the best defenses of a scorpion and black widow? A hard shell covers a scorpion's body and acts as **armor** against predators. It also keeps the scorpion safe from sandstorms and even from large amounts of **radiation**!

ARMOR KEEPS IT SAFE FROM ATTACKS

ARMOR KEEPS IT SAFE FROM SANDSTORMS

ARMOR IS STRONG ENOUGH TO PROTECT AGAINST RADIATION (SHOWN HERE IN UV LIGHT)

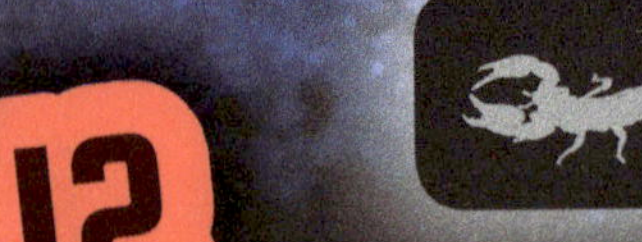

SILK IS FIVE TIMES STRONGER THAN STEEL

WEB TRAPS PREDATORS AND PREY

Black widows also have a special tool for defense: their web. They spin webs out of superstrong silk. This web creates a place to hide as well as to trap predators. A black widow's silk is strong enough to trap a scorpion. But if a scorpion is fast enough to sting the black widow, the web won't help the spider!

DEADLY VENOM

The strength of a scorpion's venom depends on its species. Most have venom only strong enough to kill insects or spiders. However, there are about 30 species of scorpions with venom strong enough to kill a human! The Indian red scorpion is the most venomous.

VENOM STRENGTH DEPENDS ON SPECIES

SOME SPECIES CAN KILL HUMANS

Black widow venom is very strong and can kill insects and small animals. However, black widow bites don't usually kill humans. Bites can cause pain, stomachache, and trouble breathing in humans. But don't think the black widow's venom isn't that scary. It turns prey's insides into liquid, so the black widow can slurp it like a milkshake!

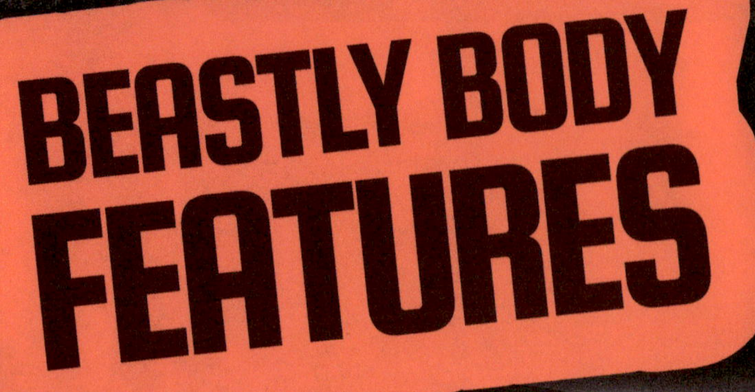

BEASTLY BODY FEATURES

Scorpions and black widows have different ways of catching prey. Scorpions have strong claws called pincers, or pedipalps, that they use like hands. The pincers crush prey or hold it in place. Scorpions have a stinger at the end of their long tail to **inject** venom into prey.

WEAPON:
STINGER TO INJECT VENOM

WEAPON:
PINCERS TO CRUSH AND HOLD PREY

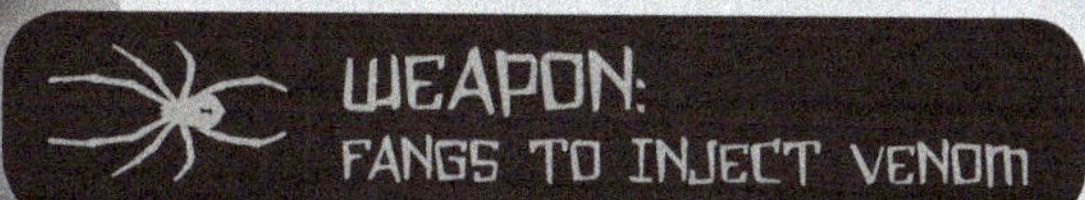

After a black widow spider's prey gets stuck in its silk web, the spider uses its special back legs with short hairs to wrap prey in silk. Then, the black widow uses its **fangs**, or chelicerae (kuh-LIH-suh-ree), to inject venom into prey. Both these arachnids have some nasty ways of dealing with prey. What do you think would happen if they used these **adaptations** against each other?

UNLUCKY PREY

Scorpions have a taste for many different creatures. They like to eat insects, lizards, and spiders, such as black widows. A scorpion might even eat a mouse. They either **paralyze** their prey with venom first or just chow down.

TASTY TREAT: LIZARDS

TASTY TREAT: INSECTS

TASTY TREAT: MICE

TASTY TREAT: ARACHNIDS

Black widows use their webs to catch flying insects, such as flies, moths, beetles, and mosquitoes. They may also eat caterpillars that crawl into their trap or grasshoppers that hop their way. They're known to eat scorpions, too. So, black widows sometimes eat scorpions, and scorpions sometimes eat black widows. This bizarre beast battle may happen more often than we thought!

AND THE WINNER IS . . .

We've compared the adaptations and behaviors of black widow spiders and scorpions. Scorpions trap prey in their deadly pincers, while black widows trap prey in their steel-strong webs. A scorpion strikes with its venomous tail, and the black widow's bite can turn guts to liquid. Both have enough venom to kill each other!

So, who would win in a fight? The outcome might change with each pair of animals! They're both deadly predators ready to kill.

THESE ARACHNIDS EACH HAVE FEATURES THAT MAKE THEM FEARSOME FIGHTERS!

GLOSSARY

adaptation: a change in a type of animal that makes it better able to live in its surroundings

armor: a thick covering or shell worn to keep someone or something safe from harm

behavior: the way an animal acts

carnivore: an animal that eats only meat

climate: the usual weather conditions in a place over a period of time

defense: a way of guarding against an enemy

fang: a mouthpart of a spider through which poison flows

habitat: the natural place where an animal or plant lives

inject: to use sharp teeth to force venom into an animal's body

paralyze: to make something lose the ability to move

radiation: harmful energy in the form of tiny particles

venom: something an animal makes in its body that can harm other animals

weapon: something used to cause someone or something injury or death

FOR MORE INFORMATION

BOOKS

Ganeri, Anita. *Scorpion.* Chicago, IL: Heinemann Library, 2011.

Randolph, Joanne. *Black Widow Spiders.* New York, NY: PowerKids Press, 2014.

Roza, Greg. *Sting! The Scorpion and Other Animals That Sting.* New York, NY: PowerKids Press, 2011.

WEBSITES

Black Widow Spider
a-z-animals.com/animals/black-widow-spider/
Read up on bizarre black widow facts.

Scorpion
animals.sandiegozoo.org/animals/scorpion
Find out more about scorpions and their fearsome features.

Publisher's note to educators and parents: Our editors have carefully reviewed these websites to ensure that they are suitable for students. Many websites change frequently, however, and we cannot guarantee that a site's future contents will continue to meet our high standards of quality and educational value. Be advised that students should be closely supervised whenever they access the Internet.

INDEX